W9-DAB-329

WITHDRAWN

INVESTIGATING
EARTH'S DESERT, GRASSLAND, AND RAINFOREST BIOMES

·ⁿ ꓷꓵꓥꓳꓕ ⁿ·
·ꓲ ꓥꓲꓵꓲ ·
ꓕꓲ ꓲꓵꓲ ·

INVESTIGATING EARTH'S DESERT, GRASSLAND, AND RAINFOREST BIOMES

EDITED BY SHERMAN HOLLAR

HUNTINGTON CITY TOWNSHIP
PUBLIC LIBRARY
255 WEST PARK DRIVE
HUNTINGTON, IN 46750

Britannica
Educational Publishing
IN ASSOCIATION WITH

ROSEN
EDUCATIONAL SERVICES

Published in 2012 by Britannica Educational Publishing
(a trademark of Encyclopædia Britannica, Inc.)
in association with Rosen Educational Services, LLC
29 East 21st Street, New York, NY 10010.

Copyright © 2012 Encyclopædia Britannica, Inc. Britannica, Encyclopædia Britannica, and the
Thistle logo are registered trademarks of Encyclopædia Britannica, Inc. All
rights reserved.

Rosen Educational Services materials copyright © 2012 Rosen Educational Services, LLC.
All rights reserved.

Distributed exclusively by Rosen Educational Services.
For a listing of additional Britannica Educational Publishing titles, call toll free (800) 237-9932.

First Edition

Britannica Educational Publishing
Michael I. Levy: Executive Editor, Encyclopædia Britannica
J.E. Luebering: Director, Core Reference Group, Encyclopædia Britannica
Adam Augustyn, Assistant Manager, Encyclopaedia Britannica

Anthony L. Green: Editor, Compton's by Britannica
Michael Anderson: Senior Editor, Compton's by Britannica
Sherman Hollar: Associate Editor, Compton's by Britannica

Marilyn L. Barton: Senior Coordinator, Production Control
Steven Bosco: Director, Editorial Technologies
Lisa S. Braucher: Senior Producer and Data Editor
Yvette Charboneau: Senior Copy Editor
Kathy Nakamura: Manager, Media Acquisition

Rosen Educational Services
Heather M. Moore Niver: Editor
Nelson Sá: Art Director
Cindy Reiman: Photography Manager
Matthew Cauli: Designer, Cover Design
Introduction by Heather M. Moore Niver

Library of Congress Cataloging-in-Publication Data

Hollar, Sherman.
 Investigating Earth's desert, grassland, and rainforest biomes / Sherman Hollar.—1st ed.
    p. cm.—(Introduction to Earth science)
 "In association with Britannica Educational Publishing, Rosen Educational Services."
 Includes bibliographical references and index.
 ISBN 978-1-61530-502-5 (library binding)
 1. Desert ecology--Juvenile literature. 2. Grassland ecology--Juvenile literature. 3. Rain forest
 ecology--Juvenile literature. 4. Biotic communities I. Title.
 QH541.5.D4H65 2012
 577—dc22

                                                              2010050464

Manufactured in the United States of America

On the cover, page 3: Africa's Sahara Desert is the world's largest desert. *Shutterstock.com*

Interior background: The sun sets on a savanna in the African country of Kenya. *Digital Vision/Getty Images*

CONTENTS

INTRODUCTION

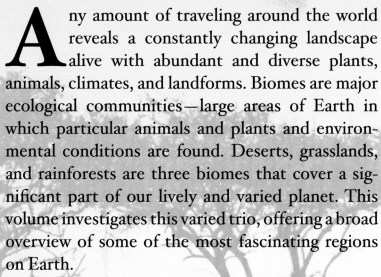

Any amount of traveling around the world reveals a constantly changing landscape alive with abundant and diverse plants, animals, climates, and landforms. Biomes are major ecological communities—large areas of Earth in which particular animals and plants and environmental conditions are found. Deserts, grasslands, and rainforests are three biomes that cover a significant part of our lively and varied planet. This volume investigates this varied trio, offering a broad overview of some of the most fascinating regions on Earth.

About 20 percent of the Earth's surface is covered by deserts. Although deserts are typically thought of as hot, they may also be cold. Most of the world's hot deserts are between 20° and 35° north and south of the equator. The main temperate (cold) deserts are found in the middle latitudes. Desert landscapes range from bare rock to boulders to sand.

Such arid areas receive scanty precipitation, and vegetation is meager. Yet little of the desert is entirely desolate. Desert-dwelling plants and animals are expressly adapted to the severe conditions of this biome. Following a good rain, plants and flowers bloom with vibrant colors. Most deserts sport grasses. North America's Sonoran Desert is spotted with the distinctive saguaro cactus *(Carnegiea*

Torrey's Yucca (Yucca Torreyi) *blooms in the Chisos Mountains of the Chihuanhuan Desert in Texas.* David Ponton/Design Pics/Getty Images

gigantea), while the cooler Mojave Desert is home to the Joshua tree *(Yucca brevifolia)*.

Deserts are home to reptiles, marsupials, and mammals. Perhaps the camel is the iconic desert animal, well suited to the climate with its ability to travel many miles without water. People also inhabit deserts and even grow crops there, though only in areas—known as oases—where water is available.

Grasslands occupy another 20 percent of Earth's land. These grass-dominated areas extend between forests and deserts. The grasses vary as soil and climate conditions change. For example, more arid areas closer to deserts grow short, or bunch, grasses. Tropical grasslands, called savannas, have a marked dry season and yield coarse, stiff grasses. In the middle latitudes, or prairies, temperatures are less extreme, and taller grasses grow alongside flowering plants in the fertile soil. Steppe grasses tend to be rather short with shallow root systems, making for a less varied landscape.

Many animals inhabit savannas, such as herbivorous (plant-eating) animals, which in turn attract carnivores, or meat-eaters. Tropical grasslands teem with zebras, lions, and kangaroos. More temperate grasslands are home to hawks, horses, and prairie dogs.

The verdant rainforest features broad-leaved evergreen trees and abundant rainfall. In many areas, the upper layer of trees creates a canopy so thick that sunlight rarely shines through to lower foliage. An incredible diversity of plants and animals is found in rainforests, such as orchids, lianas, orangutans, bats, birds of paradise, and many frog species.

Although some rainforests occur in more temperate areas, most are found in warm, tropical climes around the equator. Subtropical rainforests lie beyond the equatorial region; temperatures in these forests generally only fluctuate a little, though rainfall may be unevenly distributed. Montane and tropical seasonal rainforests diverge some from the classical definition of rainforest, but they do have periods of intense rain and moisture.

Deserts, grasslands, and rainforests certainly cover an ample portion of the globe. On the following pages, you will find a sweeping look at these captivating areas, each with its own distinctive flora and fauna.

CHAPTER 1

THE DESERT ENVIRONMENT

A biome is a large region of Earth that has a certain climate and certain types of living things. Occupying some 20 percent of the Earth's surface, deserts are one of the planet's major types of biomes. Any barren region that supports very little life may be called a desert. More commonly, however, the term *desert* is reserved for regions that are barren because they are arid, or dry. Arid deserts receive little precipitation and are characterized by specialized plants that tolerate drought conditions and salty soils. Deserts may or may not be hot. Land features range from windswept expanses of loose sand to rugged mountains, bare rock, and plateaus strewn with gravel and boulders.

CLIMATE

Deserts are areas in which there is a shortage of moisture available for plants. There is no precise measure of how dry such an area must be to be called a desert. Characteristically, deserts receive an average of less than 10 inches (25 centimeters) of rain or other precipitation per year. Many deserts get less than half that amount of rain yearly, and some receive almost none. Regions with an average annual precipitation between some 10 and 16 inches (25 and 40 centimeters) are often called semideserts.

But rainfall is not the only factor that influences how much water is available to support plants. The distribution of the rainfall throughout the year, the humidity of the air,

Desert rainfall is sparse and sporadic, so some species of cactus lack leaves, which means there is less surface area from which they can lose water. **iStockphoto/Thinkstock**

and the temperature and amount of sunshine all play a role. Some definitions of deserts focus on the imbalance between the amount of precipitation received and the amount of moisture that could be lost through evaporation and through plants' leaves. In the Sonoran Desert, in the southwestern United States and Mexico, for example, a year's potential evaporation—the amount of evaporation that would occur if water were always present—is some 100 inches (250 centimeters). This is about 20 times the actual annual rainfall.

Rainfall in deserts is not only scanty but also erratic. Most deserts have at least a few days of rain a year, but some may get no rain at all for several years or receive a year's amount in one storm. For example, Iquique, in northern Chile, had no rain for four years. The fifth year brought 0.6 inch (1.5 centimeters), making a five-year average of 0.12 inch (0.3 centimeter). At another time 2.5 inches (6.4 centimeters) fell in a single shower.

Temperatures range widely in deserts. In hot deserts, daytime air temperatures can regularly exceed 100°F (38°C) in summer. Al Aziziyah, Libya, holds the record for the hottest temperature with a high of 136°F (57.8°C), while Death Valley, Calif., comes

Rock-strewn tundra of the barren Arctic lands of Polar Bear Pass on Bathurst Island, Nunavut, Can. Regions around the poles are sometimes called cold deserts. **Brian Milne/First Light**

close with 134° F (56.7° C). Winters are cold in temperate deserts, located far from the Equator. The Gobi, in Central Asia, for example, has an average low temperature of -40° F (-40° C) in January and an average high of 113° F (45° C) in July.

The temperature drops sharply in the desert night. Dry air, cloudless skies, and bare, dry earth furnish ideal conditions for the cooling of air after sunset. A 24-hour range of 25 to 45° F (14 to 25° C) is common, and it may exceed 60 to 70° F (33 to 39° C).

WORLD DISTRIBUTION

There are two main types of arid desert: hot and temperate. Most of the Earth's deserts are found between 20° and 35° in both north and south latitudes, near the Tropic of Cancer and the Tropic of Capricorn, respectively. Except for a couple of coastal deserts kept cool by ocean currents, these deserts are hot. They occupy subtropical regions that straddle a wet, tropical region around the Equator. Because of the predominant air-circulation patterns, the air that descends into the subtropical regions has already rained out most of its moisture over the tropical lands. The air is also heated as it descends, which further prevents rain.

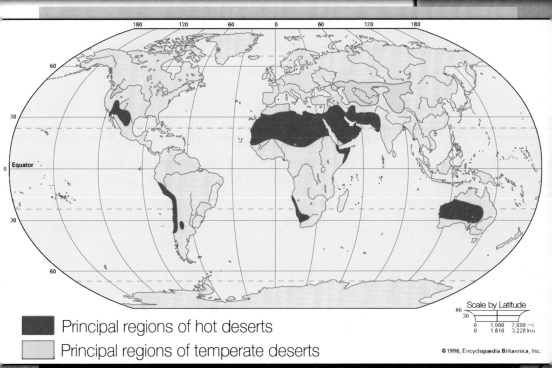

■ (dark)	Principal regions of hot deserts
□ (light)	Principal regions of temperate deserts

Scale by Latitude

60
30

0	1,000	2,000 mi
0	1,610	3,220 km

©1996, Encyclopædia Britannica, Inc.

Most hot deserts lie between 20° and 35° north and south of the equator. The main temperate deserts are found in the middle latitudes.

Temperate deserts are found farther from the Equator, in the middle latitudes. They occur mainly in Central Asia, with smaller areas in western North America, southeastern South America, and southern Australia. These deserts are generally separated from the coast by mountains or by great distance. Air picks up moisture from the ocean, but by the time it reaches these regions, it has lost much of its moisture over either the mountains or the land nearer the coast.

DEATH VALLEY

The lowest point in the Western Hemisphere, Death Valley is also famous as a scene of suffering in the gold rush of 1849. Many gold seekers nearly lost their lives in Death Valley's searing heat. They gave the valley its grim name.

The valley is a deep trough in southeastern California between the Panamint Range on the west and the steep slopes of the Amargosa Range on the east. It is about 140 miles (225 kilometers) long and 5 to 15 miles (8 to 24 kilometers) wide. Nearly 550 square miles (1,425

Death Valley.

square kilometers) of its area lie below sea level, and it contains the lowest dry land in the country, 282 feet (86 meters) below sea level. Less than 100 miles (160 kilometers) away towers Mount Whitney, 14,495 feet (4,418 meters) above sea level.

The scorching heat has reached 134° F (57° C). Despite this heat, more than 600 kinds of plants thrive there as well as rattlesnakes, lizards, coyotes, rabbits, roadrunners, and other animals. The average annual rainfall is less than 2 inches (5 centimeters). The bed of the Amargosa River, which cuts through the valley, is a series of dry channels for most of the year.

Death Valley National Monument was created in 1933. It consists of about two-thirds of the valley and a small area in Nevada. Death Valley became a national park in 1994.

Ice cap and tundra regions around the poles are sometimes called cold deserts. They have little precipitation, but the dearth of vegetation is caused chiefly by the cold.

LAND FORMS

Desert areas differ greatly in their surface features, which range from mountains to plateaus to plains. The ground may be bare rock or be covered with sand, scattered boulders,

or a "desert pavement" of coarse gravel and stones. Although sand dunes are spectacular features of deserts, they are less common than generally believed. In the deserts of the southwestern United States, for example, dunes make up less than 1 percent of the surface. In the most sandy of all deserts, the Arabian, dunes occupy only about 30 percent of the total area. If sand accumulations on plains are extensive and appear as a "sand sea," they are called ergs.

A more common type of desert consists of rugged mountains separated by basins called bolsons. The mountains receive most of the rains in downpours. As the water rushes down the steep slopes it cuts deep gullies and carries rock fragments, gravel, and sand to the bolson. These materials are freed from the water when it slows or evaporates, and they are deposited as cones or in fans of sediment called alluvial fans. The rugged forms produced in this way, such as the terrain in Death Valley, are termed badlands.

Sometimes floodwaters make a temporary shallow lake in the basin. The temporary lakes that form in basins with no outlet are called playa lakes. Typically the water soon evaporates into the air or filtrates into the ground, leaving behind sediments of clay,

Death Valley National Park is in the Great Basin region of southeast-ern California. Scenics of America/PhotoLink/Getty Images

silt, and sometimes salt. The flat-bottomed depression that is formed by the water is called a playa. In narrow basins, alluvial fans and badlands may extend to the edge of the playa. In broad valleys a surface of low relief and gentle downward slope occurs between the playa and other alluvial fans of the

mountain front. This kind of surface is called a desert flat or llano.

Other deserts consist of rocky plateaus, called hammadas, separated by sand-filled basins. Here differences in altitude are usually slight. Many hammadas are broad, almost flat, dome-shaped areas. Where streams or wind wear away the weaker rocks, strong rock formations stand out boldly as mesas or cuestas. Pinnacles, needles, and arches carved in colored rocks lend fantastic beauty to the deserts of the American Southwest. Gullies are cut deep into the hammadas by the wearing force of the torrents. Gullies are known as wadis in Arabia and arroyos in the Southwest.

CHAPTER 2
LIFE IN THE DESERT

People often think of the desert as an entirely barren place, but a desert environment hosts a community of distinctive plants and animals that are specially adapted to the harsh conditions and terrain. Although relatively few people live in desert areas, human activities have a profound impact on desert ecosystems around the world.

PLANTS AND ANIMALS

Few parts of the desert are entirely barren. Where water seeps toward the surface, a great variety of plants spring up. After a rain low shrubs and grasses come to life. At blooming time, the plants are fragrant and brightly colored. They grow quite far apart, instead of providing complete ground cover. Trees and large shrubs sometimes grow in the desert, but they are not prominent in the driest regions.

Hindu pilgrims gather at Pushkar, in the Thar Desert, Rajasthan state, India. Brian A. Vikander/West Light

Desert plants differ in the ways they are adapted to arid places. Some plants sprout when the rain falls, bloom quickly, ripen their seed in a few days, then wither and die. Other plants survive periods of water shortage by drying up and becoming physiologically inactive. When more water becomes available they swell up and resume activity. Some

Vegetation profile of a desert. **Encyclopædia Britannica, Inc.**

desert plants depend on underground water and have long root systems, while others are able to absorb dew. Various adaptations of the leaves, such as smaller surface areas, help prevent the loss of moisture. In most species of cactus, for example, leaves are either absent or extremely small. Cacti and other succulent plants also store water in their thick, fleshy tissues to help them survive long dry periods.

Animals live in all but the most barren stretches. The camel is the most useful domestic desert animal. Its physical structure permits it to travel far without water. Various wild mammals, birds, and reptiles of arid regions must get all their moisture from their food. In times of drought, many desert animals remain inactive for long periods.

The camel (genus Camelus*) is well suited to the desert's lack of water.* Bashar Shglila/Flickr/Getty Images

Others hide from the sun during the hottest daytime hours and come out at night to feed. More than half of desert animals live at least part of the time underground, where it is cooler and more humid. Another strategy is migration: many birds and other desert animals survive by regularly moving to areas that have recently received rain.

PEOPLE AND DESERTS

Deserts are much less densely populated than other land areas. People can live and grow crops in the desert only at places where they can get water, called oases. In some spots ordinary shallow wells reach the water table, or the upper boundary of underground water, but usually groundwater lies at greater depths in deserts than in humid lands. In alluvial fans the water sinks deep into the porous material, but it may be reached by a well at the tip of the fan.

In wadis, ordinary wells can usually tap a supply of good water. Oasis settlements therefore are most often found where wadis are numerous. Ergs into which many wadis drain may have a water supply. Desert shrubs in the hollows between the dunes signal its

presence. Deep artesian wells may be bored where the rock structure holds water under pressure. In some oases an artesian spring flows through a crack in the rock.

Streams that rise in rainy regions outside deserts bring the most generous supply of water for irrigation. All the large deserts except those of Australia are crossed by these so-called exotic rivers. The largest and best known of

People can live in the desert near oases, where there is water. Frans Lemmens/Photographer's Choice/Getty Images

them are the Nile in Egypt, the Tigris and Euphrates in Iraq, the Indus in Pakistan, and the Colorado in the United States.

Desert soils are usually productive when given water. They are coarse textured and highly mineralized. Most widely cultivated are the water-transported soils of flood-plains and alluvial fans. Land is precious in oases, so it is intensively cultivated. In North African and Asian oases, the chief food crops are dates, figs, wheat, barley, rice, and beans. Oasis farmers also raise such crops as cotton and sugarcane. In the United States irrigated lands are mainly used for citrus fruits, dates, winter vegetables, and cotton.

The proper development of critical water resources has focused on improving surface-water management techniques, improving control and storage of surface runoff, reducing loss from shallow aquifers, and desalting brackish waters. The buildup of salts in the soil has become a serious problem that is being studied in several nations. The introduction of cultivated plant species with greater salt tolerance, which has been done in Israel, seems to be a promising alternative to abandoning arid land.

Humans have changed many desert environments. Irrigated lands have been

extended by giant river-control systems. Oases once reached only by camel now have airports and gas stations for motor vehicles. Settlements have sprung up in deserts to obtain such valuable minerals as the petroleum of Saudi Arabia, Iraq, and Iran and the copper and other metals of arid North and South America. Typical desert minerals are soluble salts left behind by evaporation. Sodium nitrate, the most valuable, comes from the north Chile desert.

The mismanagement or overuse of arid and semiarid lands can cause desertification, or the spread of desert environments. Excessive cultivation of the land, clearing away the vegetation, or exhausting the water supply can rob arid or semiarid land of its ability to sustain life. Factors such as long-term climate change or severe drought can also cause desertification.

Cotton is one crop that is often grown near desert oases in North Africa, Asia, and the United States. **Shutterstock.com**

CHAPTER 3
GRASSLANDS

About one-fifth of the Earth's land was once covered with grass. Grasslands stretch between forests and deserts. Near the forests where rainfall is abundant, trees grow intermixed with tall grasses. As the grasslands stretch away from the forests, the rain decreases and soil conditions change. Then come stretches of treeless tall

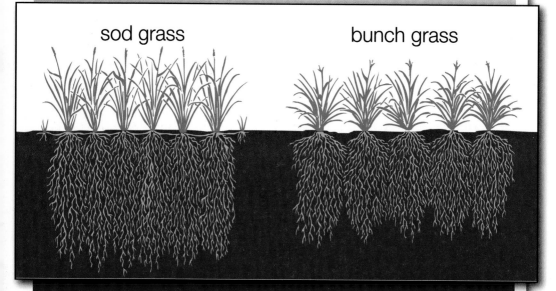

sod grass   bunch grass

Vegetation profile of a typical grassland. **Encyclopædia Britannica, Inc.**

grass. In semiarid regions near deserts, short (or bunch) grasses grow.

In low latitudes where there is a distinct dry season lie tropical grasslands called savannas. Near edges of the equatorial rainforest, trees are mixed with the grass. Gallery woods are trees that form arches over water. As rainfall diminishes, scrub forests, thorn forests, and bushes take the place of larger trees, and eventually there is only a grass cover. Where savannas border deserts, the lands are sometimes called tropical steppes.

Savanna grasses are coarse and tend to grow vigorously. They range from 2 to 12 feet (0.6 to 3.7 meters) in height. Young blades of dull green spring up rapidly at the start of the wet season. Most plants grow singly, some in thick bunches. They are separated by bare spots of reddish soil. As the plants mature, the blades grow stiff and harsh. In the dry season they change to a dusty yellow or brown and slump to the ground.

On the drier margins of savannas in Africa and Australia, the grass cover is broken by trees of the flat-topped acacia type. In the llanos of Venezuela, the campos of Brazil, and the Sudan of Africa, tall grasses are mixed with low trees and thickets.

ANIMAL LIFE OF THE SAVANNAS

Savannas are the natural home of many animals. Grass and the foliage of low trees provide food and shelter for plant-eating (herbivorous) animals. These in turn attract many

Weighing on average around 1,500 pounds (700 kilograms), the cape buffalo (Syncerus caffer) is one of the most dangerous animals on Earth. A native of the African savannas, these mammals are unpredictable when cornered, and are capable of killing a human or even a lion. **Mark Boulton—The National Audubon Society Collection/ Photo Researchers**

flesh-eating (carnivorous) animals. Although the savannas of the various continents are similar, the animal life differs widely. The South American savannas have few species of mammals, and the animals are small.

The few small mammals include red wolves, pampa deer, jaguars, tapirs, and peccaries. They do not approach the size, beauty, and majesty of the lions, leopards, zebras, giraffes, elephants, buffaloes, and other big game found on the African savannas. Mosquitoes, ants, ticks, and other insects make life miserable for animals and people of the savanna. Many birds live among the trees beside the streams, especially in South America.

Although most savannas are either plain or plateau, a few are hilly. At the beginning of the rainy season, the banks of streams are quickly flooded. In the dry seasons, the rivers return to their channels, leaving large alluvial flats to dry in the sun. The flood plains and deltas with their alluvial soils are the best places for settlements. Although savanna soils are generally better than those of the rain forests, the land is not very good either for crops or for pasture.

The small number of people who dwell on the savannas often raise stock. It is one of the

few livelihoods available for them. When the stock suffers from drought, heat, and pests, it is usually of low quality. Cattle is a means of wealth to the tribes of the African savanna. Overgrazing and grass burning can lead to serious erosion.

Some people plant gardens or fields in the rainy season. They raise sorghum, millet, yams, sesame, tobacco, and short-staple cotton.

TALL GRASSES AND RICH CROPS ON THE PRAIRIES

In the middle latitudes, with their wide range of temperatures, grasslands bear finer and shorter grasses. The prairie has tall, deep-rooted, luxuriant grasses, usually mixed with a variety of flowering plants. The grasses average 2 feet (0.6 meter) in height. A striking feature of the original prairie of the United States was the vast expanse of tall grass billowing in the wind. Except for woods along streams, the natural prairie is a treeless, rolling plain.

The prairies, in general, are in regions in which the annual rainfall averages from 20 to 40 inches (50 to 100 centimeters), with the heaviest fall in summer. In the more humid sections, there would seem to be enough rainfall for trees. Various explanations have

Grasses once covered approximately 20 percent of the Earth. © **Aurora Photos/Alamy**

been given for the complete dominance of grasses. The occasional dry years may have withered any young trees present and permitted the hardy grasses to take over. Or grass fires started by the Indians or by lightning may have killed the saplings.

Prairie soils are among the most productive on Earth. All the major prairies are today

HUNTINGTON CITY TOWNSHIP
PUBLIC LIBRARY
255 WEST PARK DRIVE
HUNTINGTON, IN 46750

important agricultural areas. When settlers came into these areas, they disturbed the balance of nature. This was especially noticeable in the United States. The settlers killed many of the native animals—deer, elk, fox, bear, bobcat, and others. They plowed up the grasses to plant crops. Some animals were wiped out, others increased. Certain birds—grouse, partridge, pheasant—were slaughtered, and animal pests, such as the gopher, increased. The chinch bug, which had fed on native grasses, attacked the farmers' grain.

STEPPES: THE GREAT PASTURES

Short, shallow-rooted grasses, often growing in bunches, cover large areas in the middle latitudes where the average annual rainfall ranges from 10 to 20 inches (25 to 50 centimeters). These are the steppes, usually located on the margins of the deserts. Mountains interrupt the pattern, so steppes do not border all dry regions. In North America, the large steppe area coincides with the Great Plains lying between the prairies and the Rocky Mountains and reaching from southern Canada into Texas.

The grasses of the steppe are usually only a few inches high. Steppe landscape is monotonous. In wetter years tall plants may rise above the grass. The best of the grassland soils are the chernozems, found on the border of prairie and steppe. They can be cultivated for long periods without using fertilizers if they are protected against erosion.

Steppes are the natural home of numerous animals, but there are not as many as on

The Pampas in Glacier National Park, Patagonia, Arg., makes up a large section of South America's grasslands. **John Eastcott and Yva Momatiuk/National Geographic Image Collection/Getty Images**

the savanna. As settlers moved in, the native animals—such as the bison, or buffalo, of the Great Plains—were slaughtered. Now people have filled nearly all the steppes with cultivated plants and domesticated animals.

Land Use in the Grasslands

The population of the world's grasslands is unevenly distributed. Most savannas and steppes—for example, the campos of Brazil and the Great Plains of North America—are thinly populated. Prairies, however, tend to be well settled. Examples include the United States Midwest and the plains of Hungary, Romania, Slovenia, and Croatia.

It was in the Old World steppe regions that most animals were domesticated. People of the Eurasian grasslands, who learned to depend on their animals, developed a nomadic, or wandering, way of life as they followed the stock from pasture to pasture. Sheep and goats could be raised best in some lands and cattle in others. Horses and camels were found useful for riding and transporting goods. Nomadic life has continued for 25 centuries, but political and economic factors make it increasingly unsuitable today.

Large machines enable farmers to harvest large fields. Shutterstock.com

Nomads

Nomads are wanderers. The word nomad comes from the Greek *nomados*, which means "wandering around in search of pasture." Today the term refers to all wandering peoples who move in cyclical or seasonal patterns during the year. Among the traditional types of nomadic groups are the hunters and gatherers and the pastoral nomads, or herders of animals.

A tribe or group of tribes who do not produce food must survive on what nature provides in the way of plants and animals. Such people cannot normally stay in one place indefinitely. When the food supply is exhausted, they move on to another source. This was the way of life for many Indian tribes of North America before Europeans arrived. Some of the San (Bushmen) of Southern Africa also are hunters and gatherers.

After the animals within walking distance of a camp have been hunted successfully and the plant food depleted, the tribe moves on. The stay in one place may be a few days, or it may be several weeks or half a year. These nomads do not wander aimlessly, however. They must know the territory in which they range: the location of the water supply, the types of plants, and the kinds and habits of game animals.

Hunters and gatherers may be said to subsist: they usually acquire enough of the

necessities to survive and no more. Because they do not produce food, they cannot provide for an expanding population. Therefore the size of the tribe remains fairly constant over long periods of time. Pastoral nomads, however, are producers of food, and the size of their tribal or ethnic units increases accordingly. These groups raise livestock, and they move about within their established territory to find good pastures for their animals.

Although nomadism of all kinds is in decline, the pastoral type persists in several parts of the world. There are pastoral nomads in Central Asia, Siberia, the Arabian Peninsula, and North Africa. Some of these groups depend on hunting as well as herding. The Urianghai of the Altai Mountains in Siberia, for example, breed reindeer, but they also depend on hunting for food. On the fringes of grasslands, families who lose some of their animals engage in agriculture from time to time, but usually as a last resort. In Southwest Asia and North and East Africa, pastoral nomadism and settled agriculture have always been interdependent. The camel-breeding Rwala Bedouin of Arabia do not engage in agriculture, but they depend on grain and other products from their neighbors in exchange for camels.

The migratory patterns of pastoral nomads depend on climate and the nature of the land. Some Kazakh groups of Central and East Asia travel hundreds of miles from winter quarters

in the south to summer pastures in the north. They carry their portable, dome-shaped tents, called yurts, as well as herd their horses, sheep, cattle, goats, and camels. Other Kazakhs travel only a few miles from winter quarters at the foot of the mountains to summer pastures at higher elevations.

In Arabia the Bedouins camp near towns or oases during the hot summer months and then move into the desert after the rainy season. They have no fixed pattern of migration as do the Kazakhs, but each group has its established territory. In southern Somalia, some groups of people live in fixed villages and send their herds out with the men twice a year— to the plateau grasslands in the rainy season and to river banks during the dry season.

Farming settlements were started centuries ago on the black prairies of Russia. Other prairie lands of the Old World have long been used for farms. In North America settlers avoided the prairies until the steel plow was invented to break the tough sod. Today the American prairies are rich agricultural and industrial regions.

Steppe areas here and elsewhere were first used as pasture by cattlemen. Settlers streamed in only after railroads had been built to carry cattle and other produce to market. Farmers succeeded ranchers as huge machines were invented to plant and harvest big grain fields. Often the farmers cultivated regions of inadequate and uncertain rainfall. In dry years winds carried away the soil in immense dust storms. Pasturelands were eroded too as overgrazing destroyed the carpet of grass.

Today efforts are being made to remedy these mistakes. Fields are returned to grass where necessary. Farmers adopt dry farming and other soil- and moisture-conserving methods. Irrigation systems are being built to supply a dependable source of water.

CHAPTER 4
RAINFORESTS OF THE WORLD

Rainforest is a term for a forest of broad-leaved evergreen trees that receives high annual rainfall and is characteristically associated with tropical and subtropical regions of the world. The broadest definition of "rainforest" also encompasses humid forests in some temperate regions. Tropical rainforest habitat is one with generally warm, equable temperatures, with those in equatorial regions typically receiving at least 5 to 8 feet (1.5 to 2.4 meters) of rain each year. Sunlight hardly penetrates the lush growth of the canopy (upper level) and subcanopies in many areas. The natural continental rainforests of Africa, South America, and Asia and those of other large landmasses such as Borneo and New Guinea have a higher diversity of plant and animal species than any other terrestrial habitats in the world. Although the different regions vary in the particular species present, the ecological processes are the same.

The upper level, or canopy, of the rainforest is so thick that sunlight hardly penetrates it. Nacivet/Photographer's Choice/Getty Images

Rainforests with thick, tangled plant growth at ground level have historically been referred to as "jungles," especially those where a significant amount of light can promote a dense growth of vines, small trees, and other plants. Jungles occur under natural situations in areas where tall trees have fallen to create open habitat that allows sunlight to enter, or along the margins of large water courses. Jungle habitat can also occur where humans have cleared rainforests of trees.

Most rainforests occur in tropical regions of the world, though temperate rainforest habitat is in certain higher latitude areas of every continent. Tropical rainforests have been defined as occurring in a band around Earth between the Tropic of Capricorn (about 23° S. latitude) and the Tropic of Cancer (about 23° N. latitude). Areas in equatorial regions generally are the most densely vegetated and have the highest biodiversity of both plants and animals.

The three major continental regions with rainforest are in 1) the enormous river basins of the Amazon and Orinoco rivers of South America and extending into Central America; 2) the central two-thirds of Africa, including basins of the Congo, Niger, and Zambezi rivers; and 3) India and Southeast Asia, including Myanmar (Burma), Thailand, and Malaysia.

Islands in the Malay Archipelago, particularly Sumatra, Borneo, New Guinea, and those of the Philippines, have significant rainforest regions, as do parts of northern Australia, most of Madagascar, and islands of the West Indies.

EQUATORIAL FORESTS

The equatorial rainforests of South America, Africa, and the Malaysian region are the densest rainforests. The largest continuous, intact rainforest is in Brazil, in a vast region

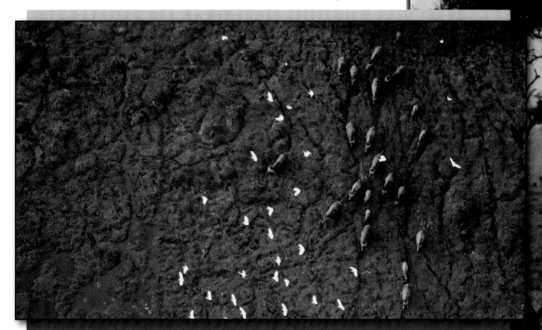

Most tropical rainforests around Africa's Congo River are in the Democratic Republic of the Congo. This aerial view shows charging buffalo and scattering egrets in Odzala National Park, Congo. Michael Nichols/National Geographic Image Collection/Getty Images

surrounding the Amazon River. More tropical rainforest has been destroyed in Brazil than in any other country, yet its intact forests account for nearly 40 percent of the tropical rainforest habitat worldwide. The second largest tract of rainforest is in equatorial Africa in the region around the Congo River. Almost half of the tropical rainforests in this tract are in the Democratic Republic of the Congo. The most extensive natural equatorial rainforests in Asia are in New Guinea and parts of Borneo.

Subtropical Forests

Subtropical rainforests are in areas outside of the strict equatorial region—but either within or bordering the tropical zone—and have more noticeable seasonal changes. Although temperatures may vary only slightly over a year, rainfall may be distributed unevenly so that wet and dry seasons occur. However, the annual rainfall is still high. Subtropical rainforests occur in Central America, the West Indies, India, Madagascar, mainland Southeast Asia, and the Philippines. In the United States, prior to widespread commercial development of southern Florida, small areas in the vicinity of the Everglades could be classed as subtropical rainforest.

HUMAN IMPACT ON RAINFORESTS

Human activities have severely disrupted the rainforests of the world. Millions of acres are lost each year to farming, logging (much of it illegal), mining, and other human endeavors. If unchecked, these activities could potentially eliminate the rainforests of most of Asia, Africa, and South America.

The largest continuous rainforest area, the Amazon, is also the largest tract of unexploited, natural rainforest. Most South American countries have some undisturbed rainforest left, and some have made efforts to protect part of what remains. In Central America, cattle ranching and cultivation have wiped out an estimated two-thirds of the region's rainforests.

Much of Africa's once vast rainforest region has been destroyed by lumbering and slash-and-burn agriculture. Slash-and-burn agriculture consists of cutting trees and other vegetation, burning what is left, and then planting crops. Because of poor soil, many such areas can support only two or three agricultural plantings before the small amount of nutrients is exhausted and the land is abandoned.

In Asia and Australia most natural rainforests have already disappeared or are

severely disrupted. Only a small proportion of the original rainforest remains on the Indian subcontinent, and most rainforests of southern China are disturbed. Logging and clearing for rubber plantations and farming eliminated two-thirds of the forests in Malaysia, Borneo, and the Philippines.

Most timbering and clearing of rainforest habitat throughout the world has occurred during the 20th and early 21st centuries. Such activities continue, yet still without an ecological understanding of the fragile nature of the rainforest ecosystems and their dependence on nutrients, which are often in short supply. Timbering and clearing activities have caused habitat fragmentation and unnatural breaks in the ecological integrity of the rainforest communities. Reforestation projects have been initiated in several countries. The previous damage to rainforest ecosystems has been extensive, however, so decades must pass before full recovery of large areas would be at all possible. Also, continued harvesting of tropical rainforests in many countries is progressing at a rate that far surpasses that at which successful recovery projects are developed. Without government officials cooperating on an international scale, in concert with ecologists, rainforests around the world will continue to disappear faster than they can recover.

TROPICAL SEASONAL FORESTS

Tropical seasonal forests receive high amounts of annual rainfall distributed unevenly throughout the year. Monsoons of the Indian Ocean region characteristically create climatic situations of heavy and continual rainfall during spring and summer, followed by a distinct dry season in fall and

A village lies within a tropical seasonal forest, or monsoon forest, in the Anaimalai Hills in southern India. Such forests occur in tropical regions that receive large amounts of rain annually but have distinct dry seasons. **Gerald Cubitt**

winter. The canopy of a monsoon rainforest is not as dense as that of an equatorial rainforest, but the lower levels are more heavily vegetated. In some regions, many of the trees are deciduous, losing their leaves during the dry seasons, and this has led to the term "semievergreen seasonal tropical forests." Monsoon forests are most prevalent in mainland Southeast Asia, Java, and northeastern Australia. Some rainforests of West Africa and South America are also affected by monsoons.

MONTANE RAINFORESTS

Montane rainforests do not meet the warm and unvarying temperature standards used to define typical rainforests. However, despite lower and wider-ranging temperatures, tropical forests in mountain regions are dense, constantly wet environments that qualify as rainforests. These forests occur in altitudes from about 3,000 feet (900 meters) to more than 5,000 feet (1,500 meters) above sea level. Some of the most notable of rainforests are in Central Africa and the New Guinea highlands.

With increasing altitude, the trees in montane forests become shorter, reaching their smallest sizes at the highest altitudes. Some high-altitude montane forests with cool, humid conditions, are referred to as cloud forests, because they are almost perpetually surrounded by cloudlike mists and fog. These forests have shorter trees and a profusion of mosses, ferns, and epiphytes, or air plants. (Epiphytes are plants that grow on trees as a substrate for support but not for nourishment.) Much of the precipitation is in the form of condensation. Cloud forests are widespread globally but are restricted to high elevation areas with moist conditions and cool temperatures.

TEMPERATE RAINFORESTS

Temperate rainforests are a distinct habitat category consisting of forests with evergreen conifers and broad-leaved trees, including deciduous species. As the name implies, temperate rainforests are outside of the tropics and bordering areas but are associated with high rainfall. The most notable are along western coastlines of large landmasses. The North American temperate rainforests are the largest in the world, ranging from northern California to Alaska and being especially prevalent in Washington and Oregon. Other

A trail cuts through a temperate rainforest in Olympic National Park, in the U.S. state of Washington. **Glen Allison/Getty Images**

regions having temperate rainforests as a consequence of oceanic rainfall patterns are parts of southeastern Australia and Tasmania, New Zealand, southern Chile, South Africa, and Japan. Western coastline portions of Turkey and Georgia on the Black Sea have also been classified as temperate rainforests, and, historically, some areas in the British Isles and Norway were placed in the category.

OTHER RAINFOREST HABITATS

A few special habitat types are considered to be rainforest variants. None of them, however, cover the broad geographic areas of the classical tropical rainforests.

The riverine forest, or gallery forest, is a special designation for tropical rainforests that follow the course of large rivers and their floodplains. Even when regional rainfall

The intertwining tree roots of the mangrove forest often create a dense junglelike habitat. Shutterstock.com

is not high, the river corridors and peripheral flooding create humid conditions characteristic of more traditional rainforests.

Another habitat often considered a modified rainforest is that of the tropical and subtropical mangrove forests that grow in the tidal zones along many coastlines and estuaries of the world. While plant species diversity is not high, and the trees are small compared to those in inland rainforests, the closeness of trees and the intertwining of their roots create an impenetrable junglelike habitat in many instances.

CHAPTER 5

LIFE IN THE RAINFOREST

The rainforest is home to a bewildering array of both plant and animal species. The vegetational structure of rainforests has been categorized in numerous ways because of the wide variety of plant community patterns observed in the forests themselves and the different ways in which ecologists have interpreted their observations. Most ecologists agree that the classic tropical rainforest has several identifiable layers composed of trees, shrubs, ferns, and herbaceous ground plants. This stratification has been described in one scheme as having five layers. First, a few extraordinarily tall trees, known as emergent trees, whose tops range from 150 to almost 200 feet (45 to 60 meters) high, extend above the tops of other trees. They are scattered throughout the forest. Second, a layer of evergreen trees form a solid vegetational cover. These trees range from 75 to 100 feet (23 to 30 meters) tall, and their tops converge horizontally to form a dense upper canopy. The third layer is an

Vegetation profile of a tropical rainforest. **Encyclopædia Britannica, Inc.**

understory stratum, or subcanopy, of trees below the canopy that are typically 45 to 75 feet (14 to 23 meters) high. Fourth, a layer of shrubs and small trees grow in the dimly lit area below the subcanopy, where most light has been blocked out by the layers of trees above. Finally, the fifth layer is a so-called ground cover of ferns and herbs.

Another characteristic of tropical rainforests is that many trees have shallow root systems and supporting buttresses as well as a network of lianas (large woody vines) that climb up, on, and over them. Many animals live in the canopy and rarely leave, directly depending on it for protection and nourishment. Epiphytic plants, including many species of orchids, ferns, and bromeliads, abound in rainforests.

Plants

The diversity of higher plant life in a rainforest is unrivaled by that in any other habitat in the world. The rainforests of the Amazon are believed to have tens of thousands of species of plants, many of them still undescribed. Because rainforests have maintained relatively uniform warm, wet climates for long periods of time, many unique terrestrial

Epiphytic orchids, which grow on trees instead of on the ground, are plentiful in tropical rainforests. **E.R. Degginger**

species of plants and animals evolved there. Often these plants have medicinal properties not found in anything anywhere else in the world. For example, quinine, used to fight malaria, and paclitaxel, used to fight some cancers, were derived from plants indigenous to rainforests. Thus, various organizations strive to preserve these rare communities of plants, because if they were destroyed, potential new medicines could be lost forever.

The large woody vines called lianas are common in tropical rainforests. Their vascular tissues are modified primarily for water conduction, which leaves these tall plants dependent on other plants for support. ©
Gary Braasch

Many plant species in rainforest canopies are not trees. Epiphytes live on tree trunks or branches and never touch the ground during their life cycle. Thousands of epiphytic species have evolved in the rainforests. Each has a complex life cycle adapted to the environmental circumstances peculiar to its particular habitat. The life histories of only a few of the epiphytic species are understood.

Dense rainforests have numerous species of vines, which have adapted in a variety of unusual ways to the minimal levels of sunlight on the forest floor. Many trees in a rainforest are almost completely covered by lianas. Some species of vines and other plants have evolved mechanisms to extract nutrients directly from the host tree. One example of the complex life cycles found among rainforest plants is that of the strangler figs whose wind-borne seeds land in the tops of trees. The strangler sends roots toward the ground from a tree, and, as the vine grows, the host tree is eventually killed by the combination of such factors as a loss of light, competition for nutrients, and physical pressure of the vine around the tree trunk.

Plants can lose nutrients by leaching when water passes over or accumulates on the surfaces of leaves and stems. In tropical rainforests with low-nutrient soils, many leaf surfaces have a waxy coating and the bark of trees is often thick, both of which reduce nutrient loss from the plant. Leaves with long, down-pointing tips increase the runoff rate from leaf surfaces.

ANIMALS

An extraordinary number of animal species exist in the rainforest, and many have complex life cycles. Because of the forest structure, many animals characteristically are associated with either the soil-litter layer, the ground surface on top of the litter, the undergrowth, the upper canopy, or the tops of the giant trees. Many species spend most of their life cycles in only one of these areas, with some species remaining in the upper canopy for their entire lives.

RELATIONSHIPS WITH PLANTS

The variety of plant forms creates a vast number of specific living places for animals, which depend on the plants for cover and for food. Through a process called coevolution, however, many rainforest plants have become dependent upon certain animals for

The wild cashew (Anacardium excelsum) grows nuts on a sweet, green stem that is eaten by bats, which disperse the nuts while feeding. Amanda Vivan/Flickr/Getty Images

pollination, seed dispersal, and protection. For example, some bats routinely obtain nectar from the flowers of certain plants, at the same time pollinating them.

Another well-known relationship is the one between acacia plants and ants. The ants raise their young inside acacia thorns by feeding their young nutritious material harvested from the plant. In return for a constant food supply, the ants patrol the tree and will attack any animal or plant that touches it. Such a relationship, called mutualism, is quite common in tropical rainforests.

Species Diversity

The numerous specific habitats available in the rainforest have resulted in the diversity of species, representing most animal groups. This is most notable among insects, but many groups of vertebrates also reach their highest species numbers in the tropical rainforests. Scientists and researchers describe new species of rainforest animals each year, but details of the life cycles and habits are known for only a relatively few species. Despite the great number of different species, the density of a single species is generally low: a common saying about the rainforest is that it is easier to collect 30 different species of a given group of plants or animals than 30 of the same kind. Social insects such as ants and termites are often the most prevalent animals in a rainforest.

Beetles, butterflies, spiders, centipedes, scorpions, and other insects and arthropods reach their greatest species diversity in the tropics. A dramatic example is that of a well-studied area of only six square miles (16 square kilometers) in Panama that has 20,000 species of insects, which is more than occur in most

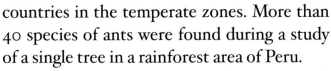
countries in the temperate zones. More than 40 species of ants were found during a study of a single tree in a rainforest area of Peru.

Among the rainforest vertebrates are such tree-oriented mammals as orangutans, gibbons, and other apes, lemurs, and numerous species of marmosets and other monkeys. Several species of cats also inhabit rainforests.

A male Raggiana bird of paradise displays its colorful feathers during a courtship display. Baiyer River Sanctuary, New Guinea; photograph, Tom McHugh/The National Audubon Society Collection/ Photo Researchers

Tropical forests harbor more species of bats than of any other type of mammal.

Bird diversity in the tropics is high, and the multitude of species include some of the most colorful in the world, such as the macaws, parrots, and birds of paradise. The rainforests of South America, Africa, and Asia are inhabited by large birds of prey (harpy eagles and African-crowned eagles, for example) that prey on tree-dwelling animals, particularly monkeys that live in the highest treetops.

BIRDS OF PARADISE

Few birds can rival the gorgeous, often bizarre plumage of the male birds of paradise. They may have iridescent neck ruffs, brightly colored, flowing tail feathers, or long, thin, wirelike projections growing from their heads. Courting males will perform for hours. In their efforts to distinguish themselves, they may wave their tail feathers over their backs or even hang upside down with their plumes cascading around them.

Some 40 species of birds of paradise are found in the forests of New Guinea and nearby islands and in Australia. They are small to medium-sized birds. Many of the males are polygamous, and the drab females generally build the nest and raise the young without the aid of the male.

Early explorers first learned of these birds when they acquired the skins from the natives of New Guinea, who used them as ornaments. Demand for the skins increased until by the early 1900s the numbers of some species had been drastically reduced. As a result it is now illegal to import wild-caught skins into the United States and most European countries.

The greater bird of paradise (Paradisaea apoda) *has been introduced into the island of Little Tobago, in Trinidad and Tobago off the coast of Venezuela.* Encyclopædia Britannica, Inc.

The best-known birds of paradise are the plume birds of the genus *Paradisaea*. Their central tail feathers look like elongated wires or twisted narrow ribbons, and their filmy tail plumes can be raised and brought forward. The blue bird of paradise *(Paradisaea rudolphi)* displays upside down, while the Emperor of Germany's bird of paradise *(P. guilielmi)* begins right side up and gradually leans forward until he is inverted.

The flag birds include the six-plumed birds of paradise of the genus *Parotia* and the King of Saxony's bird of paradise *(Pteridophora alberti)*. The latter has two 18-inch (46-centimeter) plumes trailing from the head, each bearing 30 to 40 miniature flags.

The superb bird of paradise *(Lophorina superba)* has a spreading breast shield and a broad cape that turns into a head fan. The magnificent bird of paradise *(Diphyllodes magnificus)* clears a display space for himself some 15 feet (4.5 meters) in diameter and then dances up and down in the clearing.

Other birds of paradise include the long-tailed birds of paradise (genus *Astrapia*), the manucodes *(Manucodia)*, and the riflebirds *(Ptiloris)*.

Rainforests also contain the greatest number of species of terrestrial reptiles, the snakes and lizards, on Earth. The largest snakes, the pythons of the Old World and the anacondas of South America, are characteristic rainforest forms. Among the amphibians, some attain their highest species numbers in the tropical forests. Caecilians, a group of wormlike amphibians, are represented by about 170 species that are found only in the tropical and subtropical rainforest environments. Frog species are plentiful. For example, because of its extensive rainforests, the relatively small country of Ecuador has more than 400 species of frogs, compared to only about 100 species in the United States and fewer than 30 species in France. Although tropical salamanders are restricted primarily to the Americas, frogs are common tropical species throughout the world. Reproductive behavior reaches a pinnacle of diversity among frogs in tropical rainforests.

CONCLUSION

The Earth's desert, grassland, and rainforest biomes are complex and delicate biological networks comprising numerous fragile interrelationships among plants and animals. Much of the Earth's surface is occupied by these types of biomes, and the diverse species that are adapted to them function efficiently provided that their ecosystems are not disrupted.

However, the disruption of key species (or even of a single species) in one part of an ecosystem can affect many other species and create serious environmental problems. In particular, human disturbance of these biomes—whether in the form of the removal of desert vegetation, overcultivation in grassland areas, or the timbering and clearing of rainforest habitats—can have devastating and long-term consequences. When forests are burned, for example, their carbon is returned to the atmosphere as carbon dioxide, a greenhouse gas that has the potential to alter global climate, and the trees are no

longer present to sequester more carbon. In addition, most of the planet's valuable biodiversity is found within forests, particularly tropical ones.

To appreciate the effects that the disruption or loss of species can have on these biomes, scientists must understand the relationships between organisms and their environments and the ecological processes operating within ecosystems. Further studies of these relationships and processes will help bring about a greater awareness of the fragile nature of these environments and aid conservation efforts in these regions in the future.

alluvial Sedimentary material deposited by rivers, consisting of silt, sand, clay, gravel, and organic matter.

arroyo A water-carved gully or channel.

artesian Involving, relating to, or supplied by the upward movement of water under pressure in rocks or other material beneath the Earth's surface.

badland A region marked by intricate erosional sculpturing, scanty vegetation, and fantastically formed hills—usually used in the plural form.

biome A major community of plants and animals that require similar environmental conditions.

bolson A flat-floored desert valley that drains to a playa.

brackish Somewhat salty; particularly referring to water that contains less salt than seawater but is still undrinkable.

campo A grassland plain in South America with scattered perennial herbs.

chernozem Humus-rich grassland soil used extensively for growing cereals or for raising livestock.

conifer Any of an order of mostly evergreen trees and shrubs including forms (such as pines) with true cones.

deciduous Falling off or shedding at the end of the growing period.

ecosystem An ecological unit made up of organisms and the environment they live in.

epiphyte Any plant that grows upon or is in some manner attached to another plant or object merely for physical support.

erg A desert area of shifting sand.

gully A miniature valley or gorge, originally worn in the earth by running water, through which water usually runs only after rains.

hammada A rock-floored or rock-strewn desert region.

leaching The removal of elements from the top layer of soil by percolating precipitation. The materials lost are carried downward and are generally redeposited in a lower layer.

lianas Vines rooted in the earth that climb up and wrap around other plants. Lianas are common in tropical forests.

litter The uppermost slightly decayed layer of organic matter on the forest floor.

llano An open grassy plain in Spanish America or the southwestern United States.

monsoon A large-scale wind system that seasonally reverses its direction and is frequently accompanied by heavy rainfall.

montane Of, relating to, growing in, or being the biogeographic zone of relatively moist cool upland slopes below timberline dominated by large coniferous trees.

mutualism An interaction between two species that benefits both of them.

oasis A fertile area of land that occurs in a desert wherever a permanent supply of fresh water is available.

physiology A branch of biology that is concerned with the functioning of living organisms.

pinnacle A tall, tapered, slender formation; a lofty peak.

playa The flat-floored bottom of an undrained desert basin that sometimes becomes a shallow lake.

potential evaporation The amount of evaporation that would occur if water were always present.

precipitation All liquid and solid water particles that fall from clouds and reach the cloud, including drizzle, rain, snow, ice crystals, and hail.

relief The elevations or variations of a land surface.

savanna A tropical or subtropical grassland containing scattered trees and drought-resistant undergrowth.

silt Sediment particles that are as smooth as flour when dry and hold water well. They are smaller than sand but larger than clay.

steppe Land in regions of wide temperature range (as in southeastern Europe and parts of Asia) that is dry, usually rather level, and covered with grass.

succulent A plant with fleshy and juicy tissues.

temperate Having a moderate climate that lacks extremes in temperature.

wadi The bed or valley of a stream in regions of southwestern Asia and northern Africa that is usually dry except during the rainy season and that often forms an oasis.

yurt A circular domed tent of skins or felt stretched over a collapsible lattice framework and used by pastoral peoples of inner Asia.

Canadian Ecology Centre
6905 Hwy 17, P.O. Box 430
Mattawa, ON P0H1V0
Canada
(888) 747-7577
Web site: http://www.canadianecology.ca
The Canadian Ecology Centre offers online
    research resources aimed at conserva-
    tion and development issues and options
    related to the environment as well as
    forestry and mining. Hands-on activities
    and training in outdoors skills are also
    available at the center.

Canadian Organization for Tropical
    Education and Rainforest Conservation
P.O. Box 335
Pickering, ON L1V 2R6
Canada
Web site: http://www.coterc.org
The Canadian Organization for Tropical
    Education and Rainforest Conservation
    provides leadership in education, research,
    and conservation and the educated use of
    natural resources in the tropics.

Earth Observatory
Goddard Space Flight Center

8800 Greenbelt Road
Code 130
Greenbelt, MD 20771
(301) 286-8981
Web site: http://earthobservatory.nasa.gov
The Earth Observatory features stories,
maps, images, and news that emerge
from NASA satellite missions, field
research, and climate models.

Missouri Botanical Garden
4344 Shaw Blvd.
St. Louis, MO 63110
(800) 642-8842
Web site: http://www.mbgnet.net
The Missouri Botanical Garden teaches
students about biomes such as deserts,
grasslands, and rainforests, including
the plants and animals that live there.
The botanical gardens can also be
explored online.

National Museum of Natural History
P.O. Box 37012 Smithsonian Inst.
Washington D.C., 20013-7012
Web site: http://www.mnh.si.edu
The National Museum of Natural History
is a part of the Smithsonian Institution,

a world-renowned, state-of-the-art research center and museum. The site provides students with links to exhibits, related museum Web sites, and museum collections and research information.

Passport to Knowledge
P.O. Box 1587
Morristown, NJ 07962-1587
Web site: http://passporttoknowledge.com/live.html
(973) 656-9403
Passport to Knowledge offers interactive learning about rainforests and other environments for students and teachers based on information from leading researchers.

U.S. Geological Survey (USGS)
12201 Sunrise Valley Drive
Reston, VA 20192
Web site: http://pubs.usgs.gov/gip
(888) ASK-USGS (275-8747)
The U.S. Geological Survey provides information on ecosystem and environmental health, natural hazards, natural resources, and the effects of climate and land-use change. USGS data, maps,

products, and services are available for the budding scientist interested in deserts, geology, or other related topics.

Windows to the Universe
National Earth Science Teachers Association
P.O. Box 3000
Boulder, CO 80307
(720) 328-5350
Web site: http://www.windows2universe.org/earth/desert_eco.html
Windows to the Universe has information on a variety of content for teachers and students, including news and activities to bring greater understanding about issues that affect deserts today.

World Wildlife Fund
1250 Twenty-Fourth Street NW
P.O. Box 97180
Washington, DC 20090-7180
http://www.worldwildlife.org/what/wherewework/amazon/index.html
The World Wildlife Fund works to protect the delicate ecosystem in the Amazon against such encroachments as rapid deforestation of rainforests. This site will educate users about threats to the region, wildlife, and people who live there.

WEB SITES

Due to the changing nature of Internet links, Rosen Educational Services has developed an online list of Web sites related to the subject of this book. This site is updated regularly. Please use this link to access the list:

http://www.rosenlinks.com/ies/dese

BIBLIOGRAPHY

Allaby, Michael, and Garratt, Richard. *Grasslands* (Chelsea House, 2006).

Bodden, Valerie. *Rainforests* (Creative Education, 2007).

Claybourne, Anna. *Deserts* (Smart Apple Media, 2005).

Fleisher, Paul. *Grassland Food Webs* (Lerner, 2008).

Gay, Kathlyn. *Rain Forests of the World*, 2nd ed. (ABC-CLIO, 2001).

Harris, Nathaniel. *Atlas of the World's Deserts* (Fitzroy-Dearborn, 2003).

Jackson, Kay. *Explore the Grasslands* (Capstone Press, 2007).

McLeish, Ewan. *Rain Forest Destruction* (World Almanac Library, 2007).

Oldfield, Sara. *Deserts: The Living Drylands* (MIT Press, 2004).

Sowell, John. *Desert Ecology: An Introduction to Life in the Arid Southwest* (Univ. of Utah Press, 2001).

Ward, David. *The Biology of Deserts* (Oxford Univ. Press, 2009).

Whitmore, T.C. *An Introduction to Tropical Rain Forests*, 2nd ed. (Oxford Univ. Press, 1998).